Light
Event

As written by
Mark F Kalita

1

We are all LIGHT beings.

We've just forgotten how to throw the switch.

Chapter 1
Light Formula

$$k=IH^x$$

This "Light Formula" is the k(inetic) energy that is released when the constant of the I(nfinite) is aligned by the H(uman), or LIFE COEFFICIENT. 'x' is the exponential rise of subtle energies when other H(uman)s are adjoined in the I(nfinite).

As the H(uman) becomes tuned to the potential, the k(inetic) is released. These subtle k(inetic) energies culminate in a "LIGHT EVENT" when the k(inetic) approaches the potential of the I(nfinite). This "LIGHT EVENT" shows much like the subtle electricity called static.

As the H(uman) moves closer to the I(nfinite), the potential for the LIGHT energy increases.

When released, the k(inetic) is used to serve the I(nfinite), or the UNITY OF BEING, in the form of the the THREE NATURES of existence; the Omnipresent (all present), the Omnipotent (all powerful), and the

Omniscient (all knowing).

By understanding the "Light Formula", light work becomes amplified, the cycles of life are exposed and liberation becomes available to ALL.

As this k(inetic) serves the I(nfinite), the THREE NATURES of existence are predominately revealed as HEALING. This HEALING is the process by which the positively charged I(nfinite) releases positively charged k(inetic) to further positively charge the H(uman).

When a H(uman) has full HEALING they encounter this "LIGHT EVENT" or are given the gifts of renewed youth and extension of life. The full "LIGHT EVENT" encompasses the H(uman) in the Pure Light of Being containing only peace, love and comfort.

But, more importantly, it shows how to bring the Immeasurable Light of Being, that "LIGHT EVENT" of peace and love, into this realm forever more with a small group of H(uman) beings attuned to the I(nfinite), or UNITY.

Science has, to date, ignored the H(uman) being as a measurable quotient of existence.

The "Light Formula" makes us ALL a measurable variable in the most unique equation of existence.

And the ultimate result is that H(uman) phenomenon, the "LIGHT EVENT", in which we all turn into LIGHT BEINGS of peace, love and comfort.

But, more than just the H(uman) factor, the "Light Formula" also exposes the "Unified Field Theory" in the I(nfinite) Constant and the subtle k(inetic) energies. As the "Light Formula" is at the basis of the H(uman) being, it is also the core of existence and all interaction within.

Chapter 2
I(nfinite) Constant

The I(nfinite) is ALL, the Unity of everything.

Because it is ALL, it holds the infinite potential of this ALL.

This infinite potential can only serve the ALL. As it is the universal constant in any moment, the I(nfinite) contains and supports this ALL.

Whatever is of the ALL, becomes the ALL and returns to the ALL.

This is the UNITY that the H(uman) must align their energies. As the ALL is, by nature, positively charged, the H(uman) who also charges their electrochemical process to this UNITY, or positive charge, has the resulting k(inetic) energy released to serve the ALL, or to be re-absorbed into the ALL.

It is this re-absorption process that makes the I(nfinite) the constant within this equation.

This re-absorption also explains how the other laws of constancy become variable constants within the I(finite) Constant.

It is within the moment of the reaction that the I(nfinite) has only a transfer of k(inetic) energies. No loss, only transference, thereby keeping I(nfinite) Constant.

Through this I(nfinite) Constant other physical phenomenon and physics theory are naturally wrapped into the ALL, whereby all energies effectively work kinetically and each with their own distinct "LIGHT EVENT".

Theoretically, by recognition of this "LIGHT EVENT" in other formulas, the I(nfinite) and seemingly unmeasurable is recognized as a measurable constant. Yet, measurable only in the moment of the "LIGHT EVENT".

ALL reactions of energy take place within the seemingly bubble of the I(nfinite). All theoretical processes must include this constant, and the behaviors of the constant.

As immense as this theoretical prospect is, it is also theoretically simple.

The I(nfinite) is the only constant by which all other workings depend, including the H(uman) "Light Event" of subtle k(inetic) energies.

Chapter 3
H(uman) Variable

The H(uman) electrochemical process is aligned positively or negatively to the I(nfinite). By bringing the H(uman) to be positively charged through the workings of the I(nfinite), or Unity, the H(uman) aligns with the I(nfinite) and allows the flow of the subtle k(inetic) energies within the I(nfinite).

Understanding the "Light Formula", we recognize the basis for positivity is 1 (ONE). A fraction of 1 is considered negatively charged, for in the equation the exponential factor shrinks. As the H(uman) value rises above 1, the H(uman) becomes positively tuned to the I(nfinite).

The H(uman), or LIFE COEFFICIENT, has 'Three Exposures' that comprise the positive or negative charge – ALL, WE, ME. This is the definition of our interaction in the I(nfinite), and the polarity of the charge.

$$H = (ALL)\ (WE)\ (ME)$$

The 'ALL' is the I(nfinite), or EVERYTHING.
The 'WE' is our fellow H(uman)s, or UNITY.

The 'ME' is our personal being, or BODY and MIND.

You become a positively charged H(uman) by aligning yourself with the I(nfinite). This is done by essentially cleansing, or repairing, the ALL, WE, and ME.

The 'ALL' is the I(nfinite), or EVERYTHING.

We attune the 'ALL' to the positive by living in the ONENESS of existence. By understanding the I(nfinite) as 'ALL' we learn to have a greater respect for everything that we interact with in existence.

The 'WE' is our fellow H(uman)s, or UNITY.

We attune the 'WE' when we live in UNITY. This is done by cleansing our relationships with other H(uman)s, as we ALL are in the potential of the exponent. This cleansing of the relationships, or breaking the bonds of attachment, is how we turn this aspect from negative to positively charged.

The 'ME' is our personal being, or BODY and MIND.

We cleanse our body primarily through diet and exercise. But, it is also an electrochemical process which is regulated by the subtle k(inetic) energies and their potential. The full attunement of the 'ME' involves both a care for the electro and for the chemical processes of the H(uman).

Caring for the Mind is as important as caring for the Body. As Body work is important to open up the channels of kineticism, the H(uman) must also clear the mind of delusions and cleanse the thoughts of these negatively charged aspects.

Chapter 4
k(inetic) Energy

The subtle k(inetic) energies, when positively charged appear to the H(uman) as the THREE NATURES of existence; the Omnipresent (all present), the Omnipotent (all powerful), and the Omniscient (all knowing). These energies culminate in the "Light Event". This is a similar "Light Event" that takes place with any great discharge of potential energy in a k(inetic) process within the I(nfinite).

The "Light Horizon" is when the subtle k(inetic) energies reach the I(nfinite) potential. As they are exponentially raised by the H(uman) gathered together, this "Light Event" will have a sharp curve when ALL are tuned positively, and will occur as the charge approaches the I(nfinite).

0   .0001 **<- 1 ->** 1.0001   *'LightEvent'*   I(nfinite)

Until that "Light Event", the k(inetic) energies released are very conditional and transient. Like the static, they appear as brief images and impressions of these THREE NATURES in the H(uman)s.

After the "Light Event" the positive charge stabilizes

and has a seemingly ease of permanence within the H(uman) and the proximity of this "Light Event" within the I(nfinite). As this "Light Event" stabilizes, the transference of this positively charged and stabilized k(inetic) energy becomes possible as the k(inetic), or static, qualities of this subtle energy tune the H(uman) to the positively charged k(inetic) rather than disrupt the H(uman) with the energies.

Chapter 5
Containment Body

In this "Light Formula" and the whole of the "Light Event" theory, the H(uman) is the containment structure for this "Light Event". Just as the "Theory of Relativity" shows the nuclear reaction and the containment of the nuclear reactor, the "Light Event" is the H(uman) reaction and the containment is also the H(uman) reactor. That is our design.

Even from ancient times, this design was known but could never be explained, until now. One of the most striking descriptions of this "Light Event" can be found in an ancient Judaic manuscript, the "Book of Enoch".

"And after those days my son Methuselah chose a wife for his son Lamech and she became pregnant by him and bore a son.

And his body was white like snow, and red like the flower of a rose, and the hair of his head was white like wool. And his eyes were beautiful and when he opened his eyes he made the whole house bright, like the Sun, so that the whole house was exceptionally bright."

From the "Book of Enoch"

This passage of the "Book of Enoch" describes the birth of the person called Noah in the Judaic scriptures. The "Book of Enoch" was believed to be the knowledge transmitted to Enoch, the seventh son of the person called Adam in these Judaic texts. This book was thought to be written as early as 200 BCE, if not earlier.

The description of the child in this text is a description of what the resulting containment reactor of this "Light Event" in the H(uman) may look like. While this description and explanation used the then current language and societal vernacular, with the new understanding of the 'Light Formula' we can begin to understand the workings of these subtle k(inetic) energies throughout history.

While there are many examples of these THREE NATURES in the H(uman) throughout historical texts, there is very little evidence of the H(uman) as the 'Containment Body' for the "Light Event". Whenever these events took place within our ancestors, the H(uman) reaction was explained as 'other worldly' instead of a natural process of our H(uman) body., or our LIGHT reactor.

That is our design.

That is the design of the DNA structures of the H(uman) LIGHT reactor. This design is fluid. It is able to be malleable by the healing of this containment body by the nature of the "Light Event".

As the polarity of the H(uman) moves closer to the positive, the k(inetic) energies released in the form of these   THREE NATURES naturally heal the H(uman) preparing the H(uman) to be this 'Containment Body' for the "Light Event".

Chapter 6
Transient Unity

The 'Transient Unity' process creates an electrochemical process through the endocrine system that is 'Ecstasy Based' and gives the Human a short lived connection to the I(nfinite).

These transient methods of ONENESS include:
- Dance
- Drumming / Music
- Drugs (DMT, LSD, Opiates, THC, etc.)
- Meditation / Trance

While they temporarily align the H(uman) to the ONENESS of existence, these methods make that connection fleeting and the H(uman) is always trying to re-create the experience.

In many cases, these transient methods of Unity also cause harm to the H(uman), thereby negatively charging the H(uman) within the I(nfinite). Although not irreversible in the positively charged H(uman), it does disrupt the positively charged k(inetic) flow in the H(uman).

In other cases, the 'Light Worker' will go in and out

of this ONENESS upon demand. As is the case of Light Workers who have practiced staying in this ONENESS and have achieved a level of 'Permanent Unity'.

A healer may connect with the ONENESS temporarily as the 'LIGHT WORK' is being accomplished. With practice, this healer attunes their H(uman) to the positive and this allows the k(inetic) to flow with a greater ease when the healer comes in proximity and does 'LIGHT WORK' with a H(uman) being healed.

This is the x(potential) rise in the k(inetic) when the k(inetic), or work, of the H(uman) is aligned to the I(nfinite) and contributing to the I(nfinite), or Unity. With just the healer and the other H(uman) being healed an exponent of 2 is reached. The positively charged H(uman)s create a greater healing effect for the H(uman) that is being healed.

These effects are also evident in large crowd situations in which a positively aligned H(uman) gleans subtle k(inetic) from others in the crowd, as in the case of the empathy. The crowd effect, or the x(ponential), amplifies the positively tuned H(uman). The k(inetic) fades as the crowd disperses.

Chapter 7
Permanent Unity

The 'Permanent Unity' process aligns the H(uman) through the 'Three Exposures' and gives the Human a permanence based connection to the I(nfinite).

These permanent methods of ONENESS include:
- Living in Unity as oneself
- Building Community
- Familial Bonds
- Creating Pure Lands

When a H(uman) is positively aligned to the I(nfinite) the k(inetic) begins to be revealed to the H(uman) as the THREE NATURES of existence. These have been known throughout history as the 'Greater Gifts'.

As this k(inetic) is revealed, the H(uman) finds an ease of being that allows the k(inetic) to flow effortlessly, thereby aligning the H(uman) closer to the I(nfinite) potential releasing even more k(inetic).

The "Light Event" will occur in any H(uman) that is positively charged and releases the full potential of these THREE NATURES of the I(nfinite).

The next chapter, a story from my book '7 Day Bodhi', I reveal the "Four Truths of Existence" and begin the positive tuning of the H(uman) through the understanding of the I(nfinite), or of the UNITY. This begins to build a COMMUNITY in which there can be a "Permanent Unity" process for the H(uman).

Chapter 8 is a story from another book I wrote, "Eating Free Food". "Forests of Giving" tells of a 'Pure Land' in which hunger and many of the ills of society are forever eradicated. It is within this condition of the H(uman), the 'ALL', that the 'Pure Land' concept of "Permanent Unity" has a positive permanence for the H(uman), thereby creating an ease of BEING for the k(inetic) to flow towards the "LIGHT EVENT".

Chapter 8
Unity is ALL

Without the constant and seemingly endless service that the Sun, Earth and plants provide for all other living beings, life would end.

If the Sun stopped shining, all life on Earth would eventually die off.

If the Earth stopped filtering the rays of the Sun, all life would die of the fire that would consume us.

If the plants stopped transforming the carbon dioxide into oxygen, we would not be able to breathe.

These are all the unique services that these elements give to each other to preserve life as we know it.

If our existence withheld the systems that gives us life, we all would perish.

As we are not alone within existence, we must have the awareness that we are to also serve existence and assist in upholding this harmony that insures life for all.

As servants to existence, we each remove the burdens of our unity, as we have all things in common.

As we now can live in altruistic lovingkindness, giving of yourself to other selves is only the first part of living in this awareness. The larger part is to actually be of service to others and of service to all of existence. By the measure that we offer service is the measure that we are provided service.

Our forebears were given the instruction to tend and keep the Earth. They were told to care for the fish, the animals, the birds, the plants, and all of our other selves so that we might be well.

Unfortunately these original instructions were lost for generations by the delusions of error that create suffering. Humanity has forgotten the true instructions in favor of the seemingly easy, egotistical path of greed and desire. But, those are just delusions born of the ego.

The real, easy, and liberating path of life is if all of Humanity would live a consecrated life in the unity of altruistic lovingkindness with everyone caring for all others.

This awareness corrects this error and instructs us to be of service to existence, and, in doing so, we would also be in service to all others in existence.

By tending and keeping the Earth and all of existence, we safeguard the life and well being of all others, and for all of existence. By planting herbs and trees that give us fruit and nuts, we safeguard our longevity and provide the means of liberation for all in existence.

Imagine walking amongst the Earth and, wherever you looked, you saw forests of fruit and nut trees, of every variety and age. That, when you walked through this forest of fruit and nut trees, at any time throughout the year, you only had to reach up, pick of the fruits, and eat of these gifts of existence.

Imagine that the only work that any had to do was to plant new trees, prune the ones that were growing, and gather the fruits and nuts into storehouses for times of need for all.

If our forebears had done this, hunger, and many of the ills of society, would be eliminated. All of our sufferings would have already been forgotten, as we would be living in the true awareness of existence.

Awaken to the awareness that your interaction with existence is the vehicle of liberation.

Existence does not deal in delusions, nor does it have need for awareness. Existence has always been walking in the path of liberation. You have always been walking in it too, but your awareness has been veiled in delusion.

And, it is within living in liberation that we recognize that all of the acts that we do create benefits or harm to existence, and our unity.

Being of service to others is providing, in all that we do, a level of life and well being that we would want to live.

And, we know that if everyone in existence were also concerned with abiding in this awareness, we in turn would be taken care of through the same altruistic lovingkindness that we care for all in service.

As it was spoken - The real, easy, and liberating

path of life is if all of Humanity would live a consecrated life in the unity of altruistic lovingkindness with everyone caring for all others. These are the "Four Truths of Existence":

Existence is consecrated and all are consecrated in that existence.

Unity binds existence and all are one with existence in that unity.

Altruistic lovingkindness is the core of existence as the core of all shall be.

As existence serves all, all shall give in service to existence.

Chapter 9
Forests of Giving

3745 years ago, a child was walking from a neighboring village to his village with his grandfather. He marveled at all of the fruit and nut trees that grew along the pathway and asked, "Grandfather, how did all of these wonderful fruit and nut trees come to be planted in our forests?"

His grandfather, a simple man, began to tell the story of a remarkable act of kindness, "Many, many years ago, before even my grandfather's grandfather was alive, there were many people that went hungry in our lands."

The child exclaimed, "No!"

"Yes, it is true, there was a time when our peoples had to work very hard just to be fed. That is, until a great wise man started planting fruit and nut trees every month throughout these forests. He started planting the trees when he was a young boy, about your age.

"When the wise man ate a particular fruit or nut, he would save the seeds and place them into a little pouch. Once he came upon an area he deemed necessary to plant a tree, he put the seed into the ground and tended to it until a tree started growing.

"This wise man continued this practice, every month of his long, happy life. He lived to be very old."

After pondering the generous acts of the wise old man the boy, perplexed, again spoke, "But, grandfather, that one man could not have planted all of these trees."

"No, my child," his grandfather replied.

"After the villagers saw the benefits of the wise man's actions, more and more people began to follow in his footsteps. Soon, everyone in all of the neighboring villages were planting fruit and nut trees."

"And," the grandfather continued, "still to this day, everyone in the known world understands the story of the wise man who started planting trees every month of his life, and continues in his footsteps. Our peoples know nothing else but to care for the food that was given to us, and entrusted to us, for all generations of our peoples."

The boy reached up and pulled a very ripe fig from a tree and took a very juicy bite out of it.

"Be careful, boy, I know it is hard to fight the temptation to eat our way home, but your mother has a feast readied for our return. Come, let us continue on our journey."

The child took his grandfather's hand and continued on their journey, disappearing into the fruit and nut forest started long ago by one thoughtful child.

Chapter 10
Path of Being

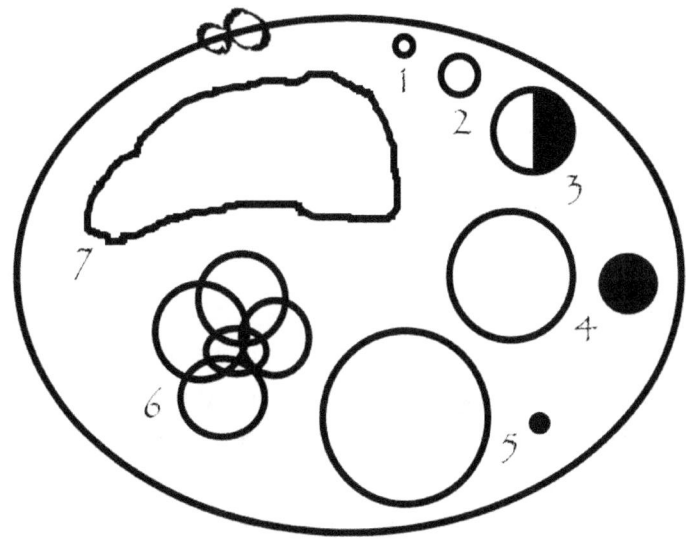

This is the path that we all take through the ALL, or UNITY of existence. This is the "Path of Being" showing how ALL experience the I(nfinite) through the path of SEPARATION. This realm is of the 3rd Octave in a struggle between positive and negative charges.

In this drawing, the OCTAVES of existence are numbered as our path through the realms of existence are shown as the "Path of Being".

The first is the SEPARATION from the ONENESS.

This is known to us as a plant or tree, an entity but not yet fully separated.

The second is the SEPARATION into BEING. This is known to us as bird, animal, or fish.

The third is the knowledge of SELF and OTHER SELF. This is known to us generally as EGOTISM or service to self and ALTRUISM or service to others. It is by this lesson in UNITY that we encounter POLARITY of BEING.

The fourth is a separation of realms into either positively or negatively charged as the EGOTISM or greed and deception and the ALTRUISM or love and compassion. The negatively charged EGOTISM collapses as the positive expands.

The fifth is a continued separation into the positive expanding and negative collapsing charges. Both are pondering the Logic and Wisdom of their path through these OCTAVES and trying to resolve their potentiality.

The sixth is the adjoining into groups of the positively charged BEINGS as the negatively charged has no course but to collapse into nothingness.

The seventh is the great adjoining of LIGHT.

The eighth is as the start and we all join hands again

-

UNITY – I(nfinite) – ALL

About the Author

Mark F. Kalita is the author of over 40 books on Spirituality and our Spiritual nature. Mark believes that it is his purpose in life to transfer this knowledge of the Ascensions and the Light of the Angelic Realm to our People.

Besides his role as Author, Mark also teaches workshops such as the "Ascension Workshop". With interests in Tarot, Crystals, Healing and other Spiritual Gifts, Mark wants to share his knowledge with others.

When Mark isn't writing or teaching, he spends time outdoors planting or planning gardens. His vision is to create a vast wilderness of food to provide sustenance for our People. From fruit bearing trees to herbs and edibles, Mark's dream is to create a new epoch of lovingkindness in which all of our basic needs are met – beginning with food!

Each new story Mark creates is an inspirational bridge to a world that could be. A world without suffering. A world bathed in the Light of knowledge and compassion founded in the Unity of Existence.

Other books by
Mark F. Kalita

These and other books can all be found at:

www.KALITA.com

Or, through Amazon in paperback or Kindle